The Sweet and Sour Animal Book

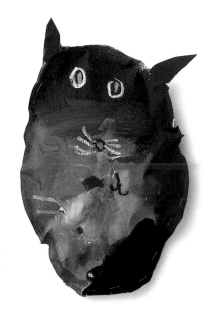

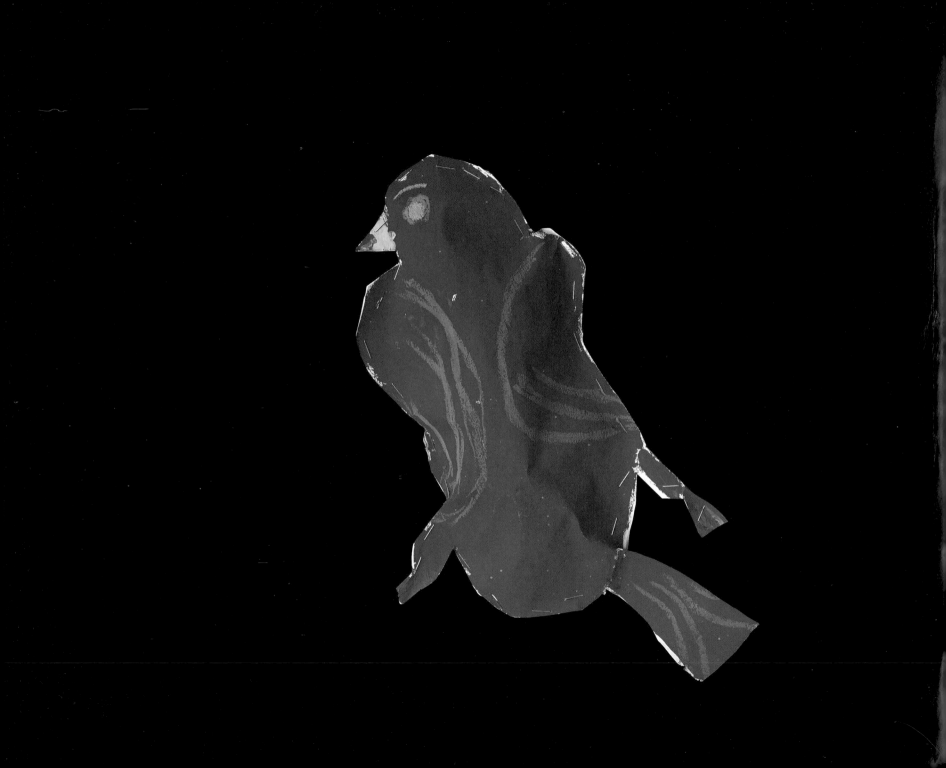

The Sweet and Sour Animal Book

Langston Hughes

Illustrations by students from the Harlem School of the Arts

Introduction by Ben Vereen

Afterword by George P. Cunningham

Oxford University Press
New York • Oxford

The Sweet and Sour Animal Book is published courtesy
of the James Weldon Johnson Memorial Collection,
Beinecke Rare Book and Manuscript Library,
Yale University.

The Harlem School of the Arts
Executive Director: Darryl Durham
President Emeritus: Betty Allen
Art department chair: David Brean
Art coordinator: Stephen Haynes
Art teachers: Lisa Bradley, Christopher Harrington,
  Jean Patrick Icart-Pierre

Design: Design Oasis
Color photography: Henry Groskinsky
Black-and-white photography: Ira N. Toff

Oxford University Press
Oxford   New York
Athens   Auckland   Bangkok   Bombay
Calcutta   Cape Town   Dar es Salaam   Delhi
Florence   Hong Kong   Istanbul   Karachi
Kuala Lumpur   Madras   Madrid   Melbourne
Mexico City   Nairobi   Paris   Singapore
Taipei   Tokyo   Toronto
and associated companies in
Berlin   Ibadan

Published by Oxford University Press, Inc.,
200 Madison Avenue, New York, New York 10016

Oxford is a registered trademark of Oxford University Press

Library of Congress Cataloging-in-Publication Information

Hughes, Langston, 1902-1967.
The sweet and sour animal book / Langston Hughes ; illustrations by students of the Harlem School of the Arts ; introduction by Ben Vereen ; afterword by George P. Cunningham.
p. cm. — (Iona and Peter Opie library of children's literature)
1. Animals—Juvenile poetry. 2. Children's poetry, American. 3. Children's
art. [1. Animals—poetry. 2. Children's poetry, American. 3. Children's art.]
I. Harlem School of the Arts. II. Title. III. Series: Iona and Peter Opie library.
PS3515.U274S94 1994
811'.52—dc20   94-8779
                                                                        CIP

ISBN 0-19-509185-X

9 8 7 6 5 4 3

Printed in Hong Kong on acid-free paper

Introduction

BEN VEREEN

The year was 1964. I was appearing in a production of *The Prodigal Son* by Langston Hughes at the Greenwich Mews Theatre in New York City. One night there were problems with the cast, and I came off stage in kind of a huff. As I walked outside there was a short gentleman, light-skinned, standing outside the theater.

He said to me, "Is this the men's dressing room?"

I said, "Yes."

"I'm looking for Ben Vereen."

"Well, what do you want with him?"

"I'm Langston Hughes."

"I'm Ben Vereen."

And he looked at me—I was kind of disheveled—and he said, "You look like you could use a dinner." And so he took me to dinner that night. That was the start of our relationship. He used to invite me up to his apartment in Harlem. He'd talk to me about his travels, and he gave me two of his books, *I Wonder As I Wander* and *The Big Sea*. We became very close at that time, though we later lost contact with each other.

It's marvelous how life comes around. Here was this man, this wonderful genius of that time—our time—who came into my life and touched me. And now, here it is years later, and Oxford University Press has rediscovered Hughes's unpublished alphabet book for children. It's a great personal privilege to be involved in its publication, to tell today's readers about my link to Langston. When we talked, he would go on and on about Harlem and how it had changed. I think he was disappointed at Harlem at that time because of all that had gone downhill, but he wasn't leaving. And the publication of this book, more than 50 years after he wrote it, with illustrations by the children of his beloved Harlem, proves that he has still not left it—and never will.

As a child, I'd never heard of Langston Hughes. When I was growing up, black history was not taught. But one thing I did have as a child was an introduction to the creative power of the arts. My godmother started me singing in church. And I had a mother who was very much like any mother. She wanted something different and better for her child. So when a talent scout came walking through the streets of our Brooklyn neighborhood looking for kids to put into a dancing school, my mother leaped at the opportunity and said, "Take my son." At first I wasn't really interested. I wanted to play stickball and chase the girls up the block. But my mother made me go to the studio, even though it wasn't very good. I guess I had a bit of a ham inside me because once I got into the classroom with all those kids, I had to rise above them. And although I didn't have the skills at that time, there was something that made me want to just perform.

One thing led to another. I'd make up dances to songs I'd heard on the radio. I found myself performing in hospitals and schools. Mr. Hill, who was the head of the glee club, saw me, and he asked me to play a bit part in the all-black production of *The King and I* at the Brooklyn Academy of Music. It was the first time I was in a show with a full orchestra, with costumes and all. It was wonderful. After that the principal got me to audition for the High School of the Performing Arts. So all along, I had some wonderful people in my life to point me in the right direction.

Performing Arts crystallized my dream. I was hooked from the first day I got to school and took my first modern dance class. It really turned my life around. I started looking at life differently, from a creative point of view. For me, the arts were great, because I wanted to be a dancer, I wanted to be an entertainer. But for any young person, the arts stimulate the creative part of the brain, and that lets you know that you can do anything, that anything is possible.

That's why I'm starting a school of the arts on the South Side of Chicago. It will begin life in a church—just as the Harlem School of the Arts did when Dorothy Maynor founded it 30 years ago. We hope to build our own facility in the near future. The curriculum will include singing, dancing, acting, and musical instruments, not only for the able-bodied but also for the physically challenged. Our job—and that of the teachers at the Harlem School of the Arts and community arts schools around the country—is to motivate children and let them know that tomorrow belongs to them.

That's also why *The Sweet and Sour Animal Book* is so wonderful. The children who created its illustrations have used their brushes and paints and clay and scissors to create a piece of the future. I believe that children come to this planet with a special gift that they will give in their time. This book gives me a powerful sense of the possibilities that await all of us.

There was an ape
Who bought a cape
To wear when he went
Downtown.

The other apes
Who had no capes,
Said, "Look at that
Stuck-up clown!"

A bumble bee flew

Right in the house

And lit on a bouquet

Of flowers.

It turned out the flowers

Were papier-mâché—

So that bee looked for honey

For hours.

There was a camel

Who had two humps.

He thought in his youth

They were wisdom bumps.

Then he learned

They were nothing but humps—

And ever since he's

Been in the dumps.

Rover Dog

Is quite brave when

He's chasing Tom Cat

Around the bend.

But when Tom Cat

Scratches him on the nose,

Rover Dog turns tail

And goes.

Elephant,
Elephant,
Big as a
House!

e They tell me
That you
Are afraid of a
Mouse.

f

There was a fish

With a greedy eye

Who darted toward

A big green fly.

Alas! That fly

Was bait on a hook!

So the fisherman took

The fish home to cook.

What use

Is a goose

Except to quackle?

If a goose

Can't quackle

She's out of whackle.

Dobbin used to be

A fire horse

Pulling a truck

With pride.

Now the village has

A motor truck—

Old Dobbin's

Cast aside.

Ibis,

In case you have not heard,

Is a long-legged

Wading-bird.

Happiest

Where fish are found,

He hates to set foot

On dry ground.

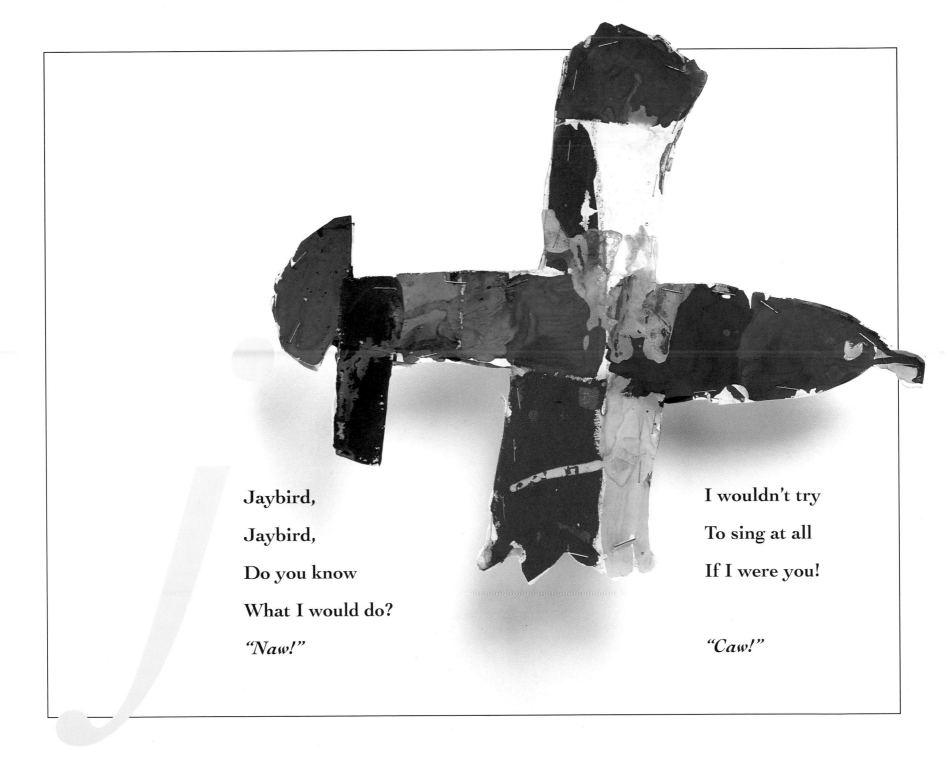

Jaybird,
Jaybird,
Do you know
What I would do?

"Naw!"

I wouldn't try
To sing at all
If I were you!

"Caw!"

A little white kitten

Got caught in the rain.

The mud and the wetting

Caused him great pain.

When he got in the house

And lay down to dry,

He started to purring,

"How happy am I!"

A lion in a zoo,

Shut up in a cage,

Lives a life

Of smothered rage.

A lion in the plain,

Roaming free,

Is happy as ever

A lion can be.

Jocko is
A peanut fiend.
He can eat peanuts
Like an eating machine.

When the peanuts are gone
And his fun is done,
Jocko can chatter
Like a son-of-a-gun!

n

Newt,

Newt, newt,

What can you be?

Just

A salamander, child,

That's me!

At night the owl
In a hollow tree,
With one eye shut,
Still can see.

But daylight changes
All of that—
By day an owl
Is blind as a bat.

There was a pigeon,
A mighty flier,
His friends all called him
Pigeon McGuire.

But he perched upon
An electric wire—
And that was the end of
Pigeon McGuire!

q

Quail

Are happy,

And fleet on their feet —

Till the hunter

Comes gunning

For something to eat!

B-O-O-M!

Peter Rabbit

Had a habit

Of eating garden plants—

Until Mrs. Rabbit

Caught Peter Rabbit

And warmed his little pants.

Mrs. Squirrel
Can look so sweet
When she finds
Her nest is neat.

When baby squirrels
Mess up her bower,
Mrs. Squirrel
Indeed looks sour.

t

Turtle, turtle,

I wonder why

Other animals

Pass you by?

Turtles travel

Very slow,

Still I get

Where I want to go.

The unicorn

Has a single horn—

Except that there is

No unicorn!

In fairy tales alone

They're born.

Happy unreal

Unicorn!

The vixen is
A female fox,
Pleased the woods
To roam.

If a trapper
Puts her in a box
She never feels
At home.

W

A pretty white mouse

Smooth as silk

Made a misstep

And fell in the milk.

When she got out

She was soaked to the skin

And mad as a hatter

Because she fell in!

X,

Of course,

Is a letter, too.

But I know *no* animal

Starts with an

X.

Do you?

Yaks are shaggy
And yaks are strong,
Happiest where
The winters are long.

But when summer sun
Is bright and bold,
A yak had rather be
Where it's cold.

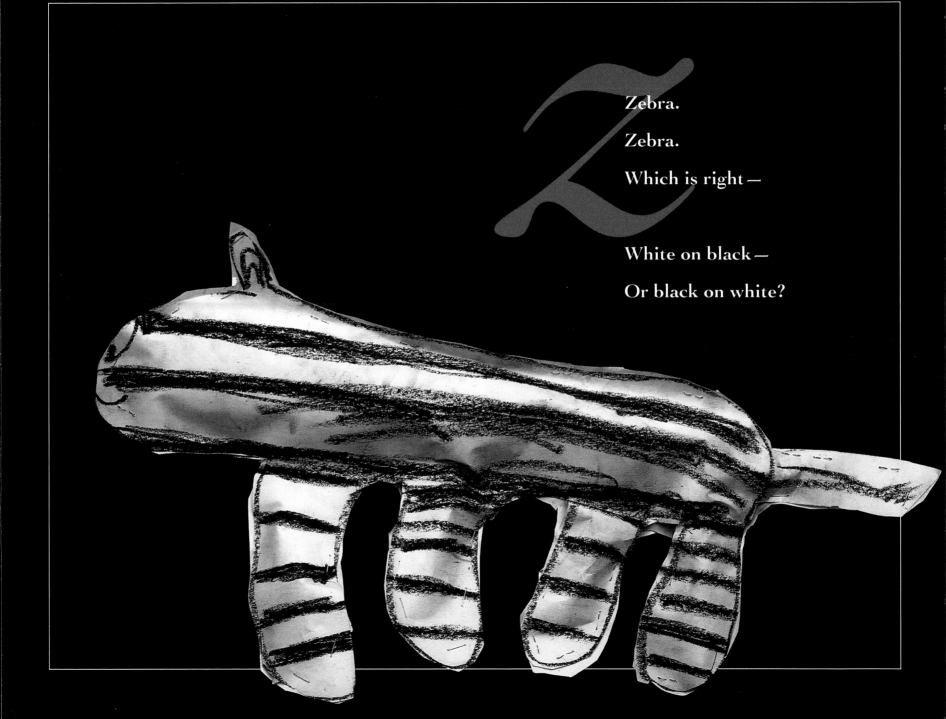

Zebra.

Zebra.

Which is right—

White on black—

Or black on white?

no more

So with a riddle,

My young friend,

From A to Z,

We come to the end.

Afterword: Serious Fun

GEORGE P. CUNNINGHAM

By the time Langston Hughes graduated from high school, he was an aspiring poet who had, by his own reckoning, written enough poetry to fill two notebooks. One notebook was "full of poems," he said, while the other was "full of verse." By verse, Hughes meant straightforward, rhythmic lines that rhymed. In the first volume of his autobiography, *The Big Sea* (1940), Hughes remarks with characteristic understatement, "I saw no harm in writing verse if you felt like it and poetry if you could." In a career that began with his first publication in 1921 and ended with his death in 1967, Hughes published more than 40 volumes of poetry, prose, and drama, establishing himself as one of the 20th-century American masters of modern literature. Yet, as much as Hughes would lead us to believe otherwise, writing what he called verse was no simple or casual matter. The complexity of one of Hughes's most famous prose characters, the black working-class figure Jesse B. Semple (Just Be Simple), should alert readers to the sophistication that Hughes brought to those moments in his writings that we take to be the most casual and even simple.

Quiet as he kept it, the verse like that in *The Sweet and Sour Animal Book* played an important role in shaping Hughes's career, his public persona, and the warm, mutually loving relationships with audiences that allowed them to embrace him as the "poet laureate of Harlem." He enjoyed and depended upon the support of the audiences that his verse brought him. In fact, throughout much of Hughes's career (he was a prodigious public lecturer) he tried to renegotiate the boundaries between light verse and seri-ous poetry. He used his verse, which was sometimes light and at other times quite serious, to expand his audience. His verse helped him to break down the barriers that walled off poetry from the everyday life of those he loved most—everyday people.

Hughes had learned the distinction between "verse" and "poetry" from his high school English teacher, Ethel Weimer, who was ahead of her time in teaching the work of the modernists Carl Sandburg, Walt Whitman, and Amy Lowell. Her own rebellion against critical dogma proved instructive. Weimer imparted to Hughes a sense of freedom in fashioning poetic forms. Thus Hughes was able, throughout his career, to test the boundaries that separated serious poetry from light verse in ways that enabled him to distinguish himself as a poet. In this regard, *The Sweet and Sour Animal Book* is best understood as a bridge between verse and poetry—or just "serious fun."

Hughes's attraction to Whitman and Sandburg was not just because they expanded the metrical possibilities of lyric voice. Their work gave him access to the prophetic power of poetry, and they inspired him to reach toward a more democratic vision of society. Hughes felt compelled to extend the range of American voices that emerged in modernist poetry. He dedicated his art to the diverse voices of African Americans in the 20th century, and he found inspiration in the rhetorical possibilities of oral performances of black Americans throughout the nation's history. Hughes's "songs of the people" drew heavily upon the collective voices of African Americans. In ways that make him radically different from Whitman and Sandburg, he masks his individuality as a poet as a means of giving voice to his poetic subjects. Reading Hughes, we are presented with the day-to-day idioms of African-American speech and song, with a lyricism that is characteristic of the blues and spirituals.

Simple rhymed poetry became an important avenue of expression for Hughes, who first came to national attention through the publication of poetry written explicitly for children. As Hughes's biographer Arnold Rampersad notes, "It was fitting, almost inevitable, that when he sought to enter the world as a poet (that is, to publish for the first time in a national magazine) Hughes offered material written from a child's point of view, or with deliberately childlike technique, to a magazine for black children." Although he had sent some of his more mature work to other national magazines, *The Brownies' Book*, published by W. E. B. Du Bois, accepted the short poem "Fairies." That first publication led to several other pieces in *The Brownies' Book* and, more importantly, to publication in *The Crisis*, the official journal of the National Association for the Advancement of Colored People. Hughes's first exposure led him to become what Rampersad calls "virtually the house poet of the most important journal in black America." His relationship with *The Crisis* during the 1920s launched his career.

By the time "Fairies" appeared in the January 1921 *Brownies' Book*, Jessie Fauset, the literary editor, had coaxed from Hughes an additional poem, "Winter Sweetness," and a brief prose piece, "Mexican Games." "Winter Sweetness" was an ideal poem for *The Brownies' Book*.

This little house is sugar.
    Its roof with snow is piled
And from its tiny window
    Peeps a maple-sugar child.

The image of the "maple-sugar child" incorporates the innocence and childlike simplicity of the fairy-tale world; but in placing a black child at the center of that world, "Winter Sweetness" accomplished the mission the editors of *The Brownies' Book* set for the magazine. The poem turns upon the metaphor of the "sweetness" of *brown* sugar as an implicit critique of the ideologies of "whiteness" that pervaded Hughes's experiences as a young African American. In the absence of anything but caricatures of African-American adults and children in the national media, Hughes, Du Bois, and Fauset wanted *The Brownies' Book* to serve the same inspirational and instructive purposes for African American children that *The Crisis* served for adults. In direct contrast to other national magazines of the time, *The Brownies' Book* was full of pictures of beautiful, well-dressed African-American children, wistful poetry, children's games, advice, and reports of individual achievement, racial news, and editorials.

With the support of *The Crisis*, Hughes became one of the most important African-American writers of the 1920s. By the end of that decade, he had published two collections of poetry, *The Weary Blues* (1926) and *Fine Clothes to the Jew* (1927), and a coming-of-age novel, *Not Without Laughter* (1930). Many of his contemporaries saw him as the representative poet of the African-American literary awakening best known as the Harlem Renaissance. Alain Locke, one of the key black proponents of the idea of a black renaissance, praised Hughes's poetry for its "Biblical simplicity of speech that is colloquial in derivation, but full of artistry." Hughes himself had helped to define the movement with his 1926 manifesto "The Negro Artist and the Racial Mountain." In it he declared: "Most of my own poems are racial in theme and treatment, derived from the life I know. In many of them I try to grasp and hold some of the meanings and rhythms of jazz."

The Sweet and Sour Animal Book was written in the 1930s, one of the bleakest periods in Hughes's professional career. The stock market crash of 1929, the Great Depression, and the loss of his chief financial patron left Hughes adrift artistically, financially, and politically. "The wolf," he wrote, "is already on Seventh Avenue." After receiving wide critical attention in the 1920s, Hughes published only one major work in the 1930s, a stunning set of short stories, *The Ways of White Folks* (1934).

Faced with the dilemma of how to make a living, Hughes wrote in the second volume of his autobiography, *I Wonder as I Wander* (1956): "If I were to live and write, at all, since I did not know how to do anything else, I had to make a living from writing itself." Much later in his career, Hughes wrote to a fellow poet about the continuing difficulties of sustaining a literary career. He gave a list of his projects, adding, "If I would just make some $ out of any one of these things, I would RETIRE and REST. But:

Money and Art
Are far apart!"

Laying the foundation for a literary career during the depression was difficult, and Hughes was torn in two directions. One direction took him, like many other American artists of the period, down the path to the left. Hughes, the most prominent African-American writer affiliated with the left, wrote poetry that became increasingly political and, as many have argued with some justice, more prosaic. *The Sweet and Sour Animal Book* stands out in contrast both to this revolutionary poetry and to the dramas of racial conflict that characterized much of Hughes's prose of the 1930s. The present volume follows Hughes's second path, one less visible in the public record. In retrospect, Hughes labeled this other path "taking poetry to the people."

This second path led Hughes to return to his origins. He published two books, *The Dream Keeper* (1932), a selection of his poetry especially for children, and *Popo and Fifina* (1932), a children's novel about Haiti coauthored with Arna Bontemps. Both works were part of a larger project of creating an African-American audience that would support a literary career. Hughes had taken so much inspiration from the day-to-day life and language of African Americans, and in the 1930s he began to look to them for something equally as important, economic support. Working with Prentiss Taylor, a white illustrator, Hughes founded the Golden Stair Press in order to publish his own work in formats that were financially in reach of a large audience. The first pamphlet from the press was *The Negro Mother and Other Dramatic Recitations* (1931), and it sold for a quarter. Aimed at a mass audience, his "dramatic recitations" consisted of what Hughes called "rhymed poems dramatizing current racial interests in simple, understandable verse, pleasing to the ear," not for the "heads of the high-brows, but for the hearts of the people." If Hughes wanted his "dramatic recitations" to reach out to a large audience, he also wanted to lure that audience to his more complex work. The central part of this ambitious plan was a lecture tour in the South. He hoped that his readings would help to build a "Negro reading public for the works of Negro authors, and at the same time, to stimulate and inspire the younger Negroes in the South toward creative literature, and the use of their own folklore, songs, and racial background as the basis for expression." Armed with *The Negro Mother* and his other works, a genial personality, and a set speech that was full of humor, Hughes set out to sell himself and his poetry to southern blacks.

Like *The Negro Mother* and the warm persona that Hughes presented in his lectures, *The Sweet and Sour Animal Book* meets his audience—in this case the youngest that he ever wrote for—halfway. This volume is one of approximately a dozen unpublished projects that Langston Hughes, sometimes in collaboration with Arna Bontemps, created especially for children. Completed in 1936, it was revised in 1952 and 1959, but was rejected over and over again by publishers. Although this little collection of light rhymed verse does not betray a racial focus, the denizens of Hughes's fanciful menagerie all seem to find their way into a blues paradox; each poem is a little existential vignette about the rewards of living.

Writing for the readers of *The Dream Keeper*, Hughes argues in "A Note on Blues" that "the blues are songs about being in the midst of trouble, friendless, hungry, disappointed in love, right here on earth. The mood of the Blues is almost always despondency, but when they are sung people laugh." Whereas Hughes was successful in refashioning the conventions of children's fairy tales and children's games in his early work in *The Brownies' Book*, this volume is shaped by a very different feeling. Though the verse is simple, Hughes infuses this collection with the same blues ethos that shaped some of his best work in the 1920s. The title perhaps leads one to expect that Hughes's animals will be alternately sweet and sour, but all are caught in the human paradoxes of life. Like the title of Hughes's 1952 collection of short stories suggests, we are left laughing to keep from crying at the frustration of the bumblebee who mistakes an artificial flower for a real one, or the paradox of the big elephant that is afraid of the little mouse.

The very lightness of the poetry eases us into more serious issues. Almost midway through his alphabetical tour, Hughes reminds us that "A lion in a zoo,/ Shut up in a cage,/ Lives a life/ Of smothered rage." Though it is clear from the rest of the poetry that Hughes does not mean to give us a thinly veiled racial drama, he is introducing young readers to a complex life, and most importantly, he offers the gift of humor as a means of transcendence. In the face of adversity, Hughes encourages us to just be ourselves.

> Newt,
> Newt, Newt,
> What can you be?
>
> Just
> A salamander, child,
> That's me!

In this volume, Hughes uses the familiar device of light verse to express profound and sometimes troubling ideas; the laughter and the ability of men and women (or animals in this case) to transcend bleak situations through laughter rings through. For the youngest readers, Hughes's menagerie is a joyful celebration of humanity. But the animals in *The Sweet and Sour Animal Book*, as in the best of Hughes's poetry, also remind us of the complexity and ambiguity of daily life. For all of us, this volume should be enjoyed as serious fun.

About this book

The Sweet and Sour Animal Book represents a remarkable collaboration between Oxford University Press and the Harlem School of the Arts. Nancy Toff, executive editor of Oxford's children's books, found the unpublished manuscript among Langston Hughes's papers at the Beinecke Rare Book Library at Yale. She and her colleagues decided that the book should be illustrated by children, rather than by a professional artist. A flutist as well as an editor, Toff was at that time organizing a collaboration between a group of professional flutists and HSA. She noticed the students' striking artwork on the cover of the school's catalog and inquired about the visual arts program. That happy discovery led her to "commission" the art department students to create their own sweet and sour animals.

And so this book celebrates not only the poetry and humor of Langston Hughes but also the creativity of children. Indeed, the project is very much in the spirit of the Harlem School of the Arts. Founded in 1964, it is the continuing legacy of the African-American mezzo-soprano Dorothy Maynor, who started the school after her retirement from the concert stage. She believed strongly that the beauty and discipline of the arts enrich the lives of all children, whether or not they are destined for artistic careers.

The school began in 1964 in a church basement, with 12 children enrolled. By 1979 the school had moved into its own architecturally distinguished building, and today the school has 59 artist teachers and provides instruction in music, dance, drama, and visual arts to more than 1,400 students a year. The school also serves the broader community through its Community and Culture program. And fittingly, the multifaceted artistic contributions of Langston Hughes have been central to many of its public programs.

Many of the children who contributed their talents to *The Sweet and Sour Animal Book*—mostly first, second, and third graders—attend HSA through Opportunities for Learning in the Arts (OLA), which provides arts instruction for more than 350 public school students during the school day. Others participate in after-school and weekend classes. The project was guided by art department chair David Brean, teacher Stephen Haynes, and their colleagues Christopher Harrington, Jean Patrick Icart-Pierre, and Lisa Bradley.

The teachers made enlarged photocopies of Langston Hughes's original manuscript and posted them in the art room. The children read the poems and chose their own animals. Unhindered by conventional scientific classifications, they let their imaginations run free. After all, why shouldn't a goose be purple? Some even wrote their own animal poems, which their classroom teachers bound into books—a process very much in concert with HSA's interdisciplinary philosophy.

Langston Hughes once wrote, "Children are not nearly as resistant to poetry as are grown-ups. In fact, small youngsters are not resistant at all." That statement is borne out eloquently by the art in this book. We think that it would please Hughes, just as it fulfills Dorothy Maynor's dreams for the children of Harlem. She said, "I want them to make beauty in this community"—and so they have.